The Quest For Finite Pi

By

Robert Charles Lewis

This book is dedicated to our Heavenly FATHER, who is recreating Himself thru all people, regardless of gender, colour of skin, faith, religion, and especially His children who are in the womb. He, our FATHER, is Above All, and beside Him always, is our Heavenly Mother.

Also by same author:

Beyond The Infinite

Poems of Multiple Meanings

Poems of Thoughtful Faith

Writings of Thoughtful Joy

Words of Truth

Developmental Peace

Truisms of Life

Continuing Life

Waterways

Hope

Tests of FAITH

Foreword

This book contains a total of 39 writings, of which 26 are poems.

The poems are placed both as a rest to the reader, and also as evidences of concepts previous to them, and even as preludes into following concepts.

When properly understood, the flow of thought is not random, but rather, continuous.

Contents

Chapter 1. Time

Concepts of Time. Thoughts About Time. More Thoughts About Time. Conclusions About Time. The Beauty of Time's Colour Rhyme.

Chapter 2. Concepts of Pi

Concepts of Finite Pi. Perimeter Pi.

Chapter 3. Poems of Time and Place

Space Travel. Stationary Movement. In Rodney Square. Caesar Rodney. Purple Applejack. Road-Kill Raccoon Stew.

Chapter 4. Central Thoughts

Chapter 5. Poems of Humanity

The Normalcy of Abnormality. Transparent Love. Past the Beyond. My Pepperflake. Venus and Mars. The Skunk-Cabbage Kid. St. Stephen's Lutheran Church. The Park of Truths. The Anatomy of Peace.

Chapter 6. Aether and Substratum

Chapter 7. A Summary

Chapter 8. Reluctances

Concept of Reluctances. A Summary of Reluctances.

Chapter 9. A Touch of Haiku

Chapter 10. Finite Pi

Chapter 1

Time

Concepts of Time

1. Time is a process.

2. Time has a pattern.

3. Time is set.

4. Time has an end.

5. Time is a perception of light.

6. Time is in the physical.

7. Time is curved.

8. Time is a relationship between the earth and the sun.

9. Time can be accelerated, stopped, and even reversed.

Thoughts About Time

1. A clock expresses time in 2 dimensions, but both could and should use 3 or more dimensions. Also, a clock can express time in 1 dimension, or even dimensionless in the physical, where time itself is a dimension.

2. "The collapse of time," in a straight physical line of 0 (zero) time or one unit of time.

3. "The completion of time."

4. "Dimensionless space."

5. "Time as a dimension."

More Thoughts About Time

1. Time squared.

2. Cubic time.

3. 4 dimensional time.

   A. 4 dimensional time expressed in 1 dimension.

4. Time of no dimension, 1 dimension, and all dimensions, expressed as 1 unit of time, or 1 unit of space, or even expressed as 1 unit of time AND 1 unit of space simultaneously.

5. Time expressed in all quadrants of a 4 quadrant graph, with time ONLY as both "scales" in each quadrant. (Therefore time can be seen as multidimensional.)

6. Time is a fixed, known constant, even in zero time or negative time.

7. Time can be expressed simply as length.

8. All geometric shapes can be "elongated" to express area as a line, whether straight or curvilinear, and thusly can be understood to be time. (A time-line).

9. Also, area in 2 dimensions can be shown with just length and time. Cubic area can be shown with length, width (or depth), and time as the third dimension.

10. Time itself is a dimension.

Conclusions About Time

Any and all shapes of space, matter, and energy, etc, can be called, simply, "elements of time," and all "elements of time" can be simply expressed mathematically as "points of time." Therefore, the present, and the future, and the past can be readily understood by simple mathematical "illustrations."

For example, the present time can be "viewed" as 0 (zero) time, and the past time can be expressed as (-) negative time, and also the future can be expressed as (+) positive time.

This then can lead readily to expressions of time containing the past and the future being ever-present, therefore a mathematical "proof" that time does not exist.

Expressing time thru the use of a quadrant, I will show some of the many aspects of time. The quadrant (s) will have a center value of zero (0) for all explanations. Also, from the center-point going "up" I will call "North" (N), and going "down" I will call "South" (S). From the center-point going to the "right" I will call "East" (E), and from the center-point going to the "left" I will call "West" (W).

Now when "North" and "East" lines always express positive values, and "South" and "West" always express negative values, this should make my examples of space (which is area), and time, clearly understood.

For example, when length is expressed on the North-South line, and width is expressed on the East-West line, then area can be illustrated.

When length is expressed on the North-South line, and time is expressed on the East-West line, then a time-line illustration is shown (on 2-dimensional paper), that shows four (4) different aspects of a time-distance relationship.

Quadrant "A" shows both time and length in the positive. Quadrant "B" shows time in the positive and length in the negative. Quadrant "C" shows both time and length as negative; and quadrant "D" shows time in the negative and length in the positive.

With this example in mind, then imagine while looking at this 2-dimensional "graph," another dimension of, say, width, or depth, or even time, coming out from the center-point, making it 3-dimensional. Also, picture a "sphere of points" emerging from the center-point of the quadrant (or any point), and it then becomes understandable that an infinite number of points can exist in 2 dimensions, or 1 dimension, and even in zero (0) dimensions.

Now, when on a quadrant both length and width are used to express area, and thereby calculating "mid-points of area" can be illustrated, then this leads to both the thought and calculations of "mid-point time." Time and length can be readily calculated and depicted, however, time can be expressed on the North-South line AND the East-West line, showing time squared, which is an area, then even in 3 dimensions, and even infinite dimensions. Any form of area, however currently difficult to express, can be shown simply as a dimension of time.

Now back to some basics of time.

A question for thought is, "What time is it at the center of the earth?" My point is, on one side of the earth at NOON, there is always a point at the opposite side of the earth where it is MIDNIGHT.

Currently, time is measured as a rotation around the earth. To me it is reasonable to assume that a center-point of time exists in the earth where both there is NO time, and even all points of time existing together mathematically.

Now using an estimate of a 24,000 mile circumference for the earth, and there being 24 hours in a day, then time, as we measure it, is moving at about 1,000 miles per hour on the "face" of the earth; and the length of its travel is NOT a straight line, but rather, a CURVILINEAR line, for the earth is not flat but it is basically a sphere.

The next question for thought is, "Does time increase in speed the further it is from a center-point of time?" For example, using our planet as a center-point of time, does an object that exists, say, 1 billion miles from earth have to move around our planet, in our 24 hour per day system of time keeping, at a very high rate of speed, to be accurately measured in earth-time, so that it is harmonious with our current concepts of time, space, and area?

An additional question is, "Is not time moving around a cube from a center-point constantly speeding and slowing relative to the "face" (meaning the outer part) of the cube?"

Likewise, any non-spherical shape would have, and express, different rates on its surface, whereas a true sphere would have a steady rate of time around its surface.

Now back to the quadrant examples:

"Time sectors" can be established and calculated using simple math, or with sine, or cosine, tangent, cotangent, secant, and cosecant formulas.

A "time sector," as I am defining it, is any area whatsoever of space, or even time alone, that exists.

These "time sectors" can show that the past, the present, and the future all exist presently, meaning now, and on the quadrant system they can be graphed and understood to be so!

Thank You Father!!!!!!!

The Beauty of Time's Colour Rhyme

Time is like colour

that differs in the present

time is a spectrum

that is sheer beauty.

As blue is to green

and green is to red

now is always the scene

where colour is in your head.

A clock is a monitor

expressing time in the present

no past nor future momentum

can be shown by present duty.

Chapter 2
Concepts of Pi

Concepts of Finite Pi

1. Definable Pi

2. Diameter Pi

3. Radius Pi

4. Center-point Pi

5. Omni-Directional Space Navigation

6. Any point or length or geometric shape Pi

7. Time expressed as Pi

8. Linear Pi

9. Curvilinear Pi

10. Dimensionless Space expressed as Pi

Perimeter Pi

1. Pi can always be converted into a measurement of a straight line, because ALL geometric shapes can be elongated into a straight line, which I call "Linear Pi."

2. A circle has both a finite area, and also can be expressed just as a clock "tells" time, which is in a curvilinear expression of portions of length around the perimeter, such as hours, minutes, seconds, etc.

3. Hourglasses express time in another way, such as a cone shape, and Pi can be arrived at as a finite value also by this thought process.

4. Time is, for the most part, a fixed constant, in all dimensions. However, both zero (0) time and negative time can be expressed to define finite Pi.

Chapter 3

Poems of Time and Place

Space Travel

The evidence is clear

that with GOD

all things are possible.

So we need to be

in a state of agree

with every truth of HE.

Eventually

mankind will see

a way to "ski"

thru-out eternity.

With perfect technology

we can build easily

vehicles that defy

the logic of humanity.

Space travel will be

like a walk in the park

in an instant or more slowly

where you want to go you will be.

Stationary Movement

Being still

in one place

is a thrill

when minds race.

With a thought

we can be anywhere

past or present

without scare.

Imagine creation

of the earth

it's a sensation

now you are bought.

From beginning to end

this message I send

our mind is eternal

so give it a whirl.

In Rodney Square (A Song)

A pretty green lawn

on a cloudy day

the comfort makes me yawn

this I sing and say.

Father above

thank you for this love

pigeons of the sky

don't seem to fly.

A cool and gentle breeze

rattle the leaves

trees clap hands this way

which causes me to pray.

Our King Christ Jesus

always You please us!

Caesar Rodney

Though a statue I see

of thee on a steed

there is really no need

for this memory.

To "Philly" you rode

to cast a vote

but by your coat

you look like a toad.

How many a bog

did you cross on your way?

Your horse needed hay

you belong on a log.

Purple Applejack (A Song)

"Purple Applejack"

nothing it does lack

this you will see

for Kosher it shall be.

Grown in every vale

this future of our land

certainly is most grand

this brew is a sure sale.

With "Purple Applejack"

you'll sleep sound in sack

this guarantee I'll back.

Road-Kill Raccoon Stew

On North Franklin Street

near the "Bennett House"

lay a dead raccoon

from "Cool Spring" Park.

Now to some it's a treat

in spite of louse

for they eat with spoon

this critter of the dark.

I went home to get bags

to carry him to

the nearest trash bin

but upon my return

I knew his tail was a-wags

cooking in a stew

to me it's a sin

to eat such "burn."

It was plain to see

that his carcass was gone

to a house that stank

and that he "swam" in a brew

most people would agree

and would hurry on

past a smell so rank

coming from "road-kill raccoon stew."

However, I still have grins

of this foul "spoon"

from Dr. James Tilton's

blue lagoon.

Chapter 4
Central Thoughts

Central Thoughts

1. All shapes, even infinite in dimension, can be defined and expressed in the finite.

2. Time, also, can be expressed finitely, even as a dimension or as an area.

3. All areas and dimensions can simply be expressed as 1 UNIT.

4. Area can be expressed in ZERO time (non-existent time), and also in plurals of time, meaning INCREMENTS OF TIME.

5. The earth can be expressed as 1 UNIT of area, also, the universe can be expressed as 1 UNIT of area, and lastly, "outer space" (which is both infinite and even beyond the infinite) can simply be expressed as 1 UNIT (of both area and time).

Chapter 5

Poems of Humanity

The Normalcy Of Abnormality

Unique we all be

as far as I see,

so as for abnormality

in truth it is normalcy.

If you think this is not so

then I will simply show

whether amateur or pro

a fact to make you know.

Consider our D.N.A.,

and with this I say

identity is known this way,

case closed, ole!

Transparent Love

Love that is evident

Shows thru the eyes

and is truly meant

nor wears a disguise.

Love starts with a parent

and transmits to a child

though clothes may be rent

love is always mild.

Transparent love therefore

will make us adore

Our FATHER of light

for we are in His sight.

Past The Beyond

Past the Beyond

on a golden street

by a platinum pond

there we will meet.

Most beautiful city

far, far above

you my kitty

my dove of love

with teeth of pearls

I enter thru

your gates of love

with many girls

for I AM a Jew

yet you are my True Dove.

Jerusalem my mother

there is no other

though with my many-a-kitty

I will live past eternity.

My Pepperflake

My little pepperflake

how the sun do bake

you on a hot day

but don't fly away.

Speckled you surely be

to hide in a tree

from all your prey

it's our Creator's way.

Oh apple of my eye

your beauty is so sly

for birdie thou may be

our Father's hand I see.

Venus and Mars

Women are from Venus

and men are from Mars

this is so because

we're each others candy bars.

Now a Venus flytrap

is a beckoning sight

it is able to zap

without even a bite.

A clue to my tale

is Jonah in a whale

as he was saved by grace

so is the human race.

The Skunk-Cabbage Kid

Referring to some food

it's not a certain mood

but foul some be

when my nose does agree.

The smell of a platter

is the chief matter

though it appeals to eye

some food is not for I.

Should my nose fail

my taste buds rule

for they are no fool

this way I don't ail.

Venus and Mars

Women are from Venus

and men are from Mars

this is so because

we're each others candy bars.

Now a Venus flytrap

is a beckoning sight

it is able to zap

without even a bite.

A clue to my tale

is Jonah in a whale

as he was saved by grace

so is the human race.

The Skunk-Cabbage Kid

Referring to some food

it's not a certain mood

but foul some be

when my nose does agree.

The smell of a platter

is the chief matter

though it appeals to eye

some food is not for I.

Should my nose fail

my taste buds rule

for they are no fool

this way I don't ail.

St. Stephen's Lutheran Church

In Heaven I am sure

Saint Stephen is glad

for thru this church door

comes free food in times bad.

Though first martyr he be

I know he sees me

for Heaven he saw

when testifying of all.

Jesus Christ led his way

now this I do say

a servant I'll be

to Saint Stephen with glee.

The Park Of Truths

Freedom does ring

my heart always sings

in memory of our King

where reality stings.

Many have fallen

in giving their all

but Christ has risen

and hears their call.

Three round pools

like a gentle rain

are memory schools

to lessen the pain.

The Anatomy Of Peace

How will peace come?

for it is wholesome

all inner turmoil

causes blood to boil.

Diffuse the heat

dose by dose

then we will defeat

all our woes.

Chapter 6

Aether and Substratum

Concepts of Aether and Substratum

1. The "Aether" is a stationary energy that exists in all space.

2. "Substratum" are energies that exist additionally in space.

3. The "Aether" alone is the foundation of all other energies.

4. "Substratum" are infinite in number.

5. "Substratum" are energies that can move thru the "Aether," and can convey infinite forms of energy to any and all points in the "Aether."

6. The "Aether" is in constant communication with the "substratum."

7. Unlimited energies and mass can be brought into existence at any point in space instantly thru the workings of the "Aether" and "Substratum," and likewise can be taken out of existence instantly.

NOTE: Biblical evidences lead to support the existence of Aether and Substratum.

For example, light was created instantly, even before the sun and stars came into existence, AND light can be turned on and off instantly at times.

Also, it is very possible that the sun and all the stars are being "fueled" in place, in order for them to either continuously or "randomly" emit various forms of energies.

It becomes more and more clear that the visible, as well as the currently detectable, were made and are sustained by the invisible, meaning powers yet unknown.

The existence of creative power should be accepted.

Chapter 7

A Summary

A Summary

Mathematics of all types are simply a language of symbols, exactly as is the alphabet.

Geometry, algebra, trigonometry, etc, merely use additional symbols of communication.

Therefore, ALL math inclusively, along with all languages, I call the "OMEGABET."

Chapter 8

Reluctances

Concepts of Reluctances

A. Diagonal Opposition

B. Rotational Diagonal Opposition

C. Cantilever Rotation

D. Cantilever Rotational Time

E. Tandem Rotation

F. Tandem/Cantilevered Rotation

G. Tandemly Rotating Time

H. Tandemly Rotating Gravity

I. Seasonal Time

J. Gravitational Reluctance(s)

A Summary of Reluctances

The term "Reluctance" I will define simply as the overall effects of any change in the universe.

Now, I will classify this term into two (2) main categories, being:

1. "Offsetting Reluctance," and

2. "Non-Offsetting Reluctance."

"Offsetting Reluctance" is when objects in the universe remain in a seemingly perpetual state of uniformed and predictable pattern of events.

"Non-Offsetting Reluctance" is when objects in the universe *do not* remain in the "Offsetting Reluctance" category. In times of "Non-Offsetting Reluctance," the effects on the universe, in various locations, can be mild to catastrophic.

As an example, a pond on earth can be completely calm, but when a pebble is tossed into it, the water has small ripples of waves. Likewise, the Pacific Ocean can change in an instant from calm water to an incoming one hundred foot (100') tidal wave of extremely destructive power, just by a relatively small shift of the ocean floor height, regardless of cause.

Now, as gravitational flows of energies change relative to the earth, these energies can have both a calming effect on the various forces at work on the earth and its atmosphere, and likewise can cause a disruptive effect on the earth. Reportedly, the moon is moving nearer to the earth, and this alone can account for climate change and "roaring seas."

As I discussed in my first (1st) book, titled *Beyond The Infinite*, gravity itself sets the bounds of all motions of the heavenly bodies. Ergo, the moon may be nearing the earth, but gravity will pull it away from the earth, just like clockwork.

Seasons on earth are NOT fixed amounts of time as we currently measure them. Rather, spring may be two(2) months in duration, followed by a five (5) month summer, a one (1) month autumn, and a (4) four month winter. However, a calendar year need not be twelve (12) months, but can be other ranges of time (for example, 10 months or 13 months). Time itself is not a fixed unit on earth, even evidenced by a volcanic eruption changing time by one (1) second.

Rest assured that any period of "global warming" will ALWAYS be followed by a period of "global cooling." Our planet will endure, as well as human life, even beyond the end of time.

Our sun is also affected by "Reluctance," and this explains varying amounts of solar energy outbursts, which in turn affect our climate, health, crops, and joy.

Planets DO HAVE orbital fluctuations, and coupled with incoming and outgoing gravitational winds of energies, any perceived change is only in the short term, NOT in the long term overall balance of continuity of life.

Chapter 9

A Touch of Haiku

Haiku

bottleneck pass

deer trail

venison

                                                            fresh water pond

                                                            deep water

                                                            fish

dinosaur bones

cliff-side

ancient

                                                            passage of time

                                                            shifting sand

                                                            crude oil

cascading water

fresh mist

waterfall

Haiku

trumpet vine

summertime

hummingbird

        scattered clouds

        sunshine rain

        rainbow

warming soil

gentle rains

spring time

        farmers plant

        annual crops

        new year

thunderstorms

heavy rains

drinking water

Chapter 10
Finite Pi

Finite Pi

There is no need for an inexpressible concept of the area of a circle, namely Pi, nor in any other geometric shape, because all shapes are definable simply as a set of points, meaning a sum of points.

Do mathematicians need to resort to, say, complex equations, such as Bases of Abstract 1, Abstract 2, Abstract 3, etc, to arrive at this conclusion?

In some instances, length can be expressed exactly with the use of inches (which is a form of Base 13), in combination with the metric system (which is a form of Base 10). Time can be shown in a 24 hour day to be Base 25; hours are expressed in Base 13, minutes in Base 61, etc (Time, ergo, can be expressed as a multifaceted dimension with these concepts). Also, the expression of "ZERO" (-0-) can be proven to have a negative value, a positive value (as do all other numbers currently existent in our system of math); and time can also be shown to have a NEUTRAL value, meaning, and even proving that time can "stand still"; and with negative time that means time travel, etc, into the past is possible, as well as positive time showing that time travel into the future is possible, which is now self-evident.

May those inclined to do so, work out the equations.

GOD BLESS ALL.

Byword

Isaiah 53:3

    Jesus Christ was not welcomed on earth.

Romans 15:6

    All of us should honour our Heavenly Father.

Numbers 28:4

    One lamb was slain in the morning, and one lamb was slain in the evening.

I Corinthians 15:2

    Remember what you were taught, and stand on those truths.

I Chronicles 13:1

    King David took counsel.

III John 11

    Always do good, and never do evil.

Revelation 5:12

    The lamb that was slain is worthy to receive all goodnesses.

May those inclined to do so, work out the equations.

GOD BLESS ALL.

Byword

Isaiah 53:3

    Jesus Christ was not welcomed on earth.

Romans 15:6

    All of us should honour our Heavenly Father.

Numbers 28:4

    One lamb was slain in the morning, and one lamb was slain in the evening.

I Corinthians 15:2

    Remember what you were taught, and stand on those truths.

I Chronicles 13:1

    King David took counsel.

III John 11

    Always do good, and never do evil.

Revelation 5:12

    The lamb that was slain is worthy to receive all goodnesses.

Genesis 1:26

    God said, Let us make man in our image.

John 1:1-17

    Jesus Christ is God's Word made flesh.

Luke 24:49

    The lamb slain at evening is the Holy Spirit made flesh, namely, the Comforter.